Plants Grow

Characters

 Red Group

 Blue Group

 Purple Group

 All

Setting   Different gardens

My Picture Words

 **apple**

 **cucumber**

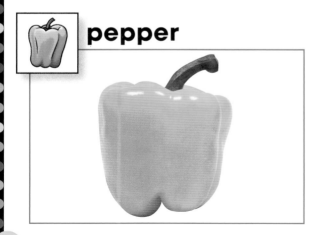 **pepper**

 **watermelon**

My Sight Words

at

can

eat

look

the

was

Look at the  .

apple

 The  was a .

apple   seed

 We can eat the !

apple

4

Look at the  .

cucumber

 The  was a .

cucumber   seed

 We can eat the  !

cucumber

Look at the  .

pepper

 The  was a .

pepper   seed

 We can eat the !

pepper

Look at the  .

watermelon

 The  was a  .

watermelon   seed

 We can eat the  .

watermelon

Look at the  .

apple

 Look at the  .

cucumber

 Look at the  .

pepper

 Look at the  .

watermelon

The End